STAR COPPER COMPANY.

REPORT

OF THE

DIRECTORS AND MINE AGENT

OF THE

STAR COPPER COMPANY,

WITH

Letters from Wm. H. Stevens and A. C. Davis, Esqs.

BOSTON:
PRINTED AT THE HERALD JOB OFFICE, 4 WILLIAMS COURT.
1864.

Star Copper Company.

TWENTY THOUSAND SHARES.

Incorporated under the General Mining Laws of the State of Michigan.

OFFICERS.

PRESIDENT.

HENRY CROCKER.

SECRETARY AND TREASURER.

H. W. NELSON.

DIRECTORS.

HENRY CROCKER, - - - -	Boston.
A. W. SPENCER, - - - -	"
CHARLES D. HEAD, - - -	"
FARNHAM PLUMMER, - - -	"
FREDERICK NICKERSON, - -	"
H. W. NELSON, - - - - -	"
HENRY BUZZO, - - - -	Michigan.

STAR COPPER COMPANY.

The location of the STAR COPPER COMPANY is on Keweenaw Point, Lake Superior, Michigan. It was purchased in 1850, by persons residing in the mining district, who had every opportunity for making a good selection of mining land, and was secured for the purpose of mining for copper. It is composed of the following described tracts of mineral land, viz: South half of the North-East quarter of Section Nine; South East quarter of Section Nine; North half of Section Ten; North-West quarter of North-West quarter of Section Fifteen, in Township Fifty-eight North, of Range Twenty-eight West,—in all six hundred acres.

A company was organized on the Lake, and explorations made, which exposed several well-defined copper-bearing veins, and regular mining was commenced.

In 1851 a controlling interest of the stock was purchased by residents of Boston, and the principal office removed to this place. About seventy thousand dollars have been expended in opening the veins on the location.

On the principal vein quite a mine has been opened, two shafts having been sunk, one of which is more than three hundred feet deep; the other down to the thirty-fathom level. These shafts are connected by galleries, and an adit level for drainage, which level will strike the green stone at nearly two hundred feet in depth from the surface.

The vein is large, well defined, and well filled with copper. But little stoping has been done in the mine. From the very first opening on this vein its appearance has been promising, and the property has been highly valued by those best qualified to judge of its worth. All the work that has been done has but increased the confidence in the ultimate success of this mine.

The financial revulsion of 1857 caused suspension of the work; but the stockholders who first invested in this property are stockholders still, and the mine is favorably regarded by them.

The majority of the stock, however, has recently been purchased by persons thoroughly acquainted with mining operations, and with long experience in those of Lake Superior.

During the past winter a tract of forty acres of land has been purchased by the company, through which runs the Montreal River, which will not only yield all the water necessary for stamps, but will also furnish a water power sufficient, it is thought, to crush one hundred tons of vein rock per day.

This water power is of very great value, as it will obviate the expense of costly steam engines, engineers, fuel, and other expenses incidental to the working of all mines now wrought with success on Lake Superior.

The property of the Company has precisely the same geological formation as the Cliff, Central, Amygdaloid, Pennsylvania, and other valuable mines, as will be seen by the accompanying map and section; and it is probable that when the vein is opened underneath the green stone, and on the same belts, that heavy mass copper will be found.

The mine is less than three miles from Copper Harbor, one of the best harbors upon the Lake, to which there is an excellent road. Upon the land is every variety of timber for fuel, building, and mining purposes.

There are several buildings upon the property, a steam engine, and a large pile of vein rock ready for the stamps.

An amygdaloid belt, carrying copper, thirteen feet in width, has been opened about one hundred and fifty feet east of the transverse vein, and it appears richer here than at any other place yet opened upon the surface on Point Keweenaw. This belt can be drained and worked in depth from the mine.

A cross cut east from the mine at the third level has been driven, the end of which is supposed to be near another transverse vein, which carries rich barrel work where it has been opened on the surface.

Sixteen heads of Gates' Patent Rotary Stamps have been ordered, which will be erected to run by water power as soon as possible. A tram road will be built from the mine to the mill, to obviate the hauling or unnecessary handling of stamp work ; and it is thought the mine can be opened by the time the mill can be put into operation, so as to furnish vein rock sufficient to keep the stamps running.

There are at present over twenty thousand dollars in the treasury, and the Company is without debt.

The letters of William H. Stevens and A. C. Davis, Esqs., annexed, give the opinion of our best mining men upon the worth of the Star property.

The mining will be vigorously prosecuted, and taking everything into consideration, it is fully believed that there are few chances so certain of profit and so guarded against loss.

Energetic measures have been taken to bring the value of the mine into that position which it should justly hold.

HENRY CROCKER, A. W. SPENCER, CHARLES D. HEAD, F. NICKERSON, FARNHAM PLUMMER, H. W. NELSON, HENRY BUZZO,	*Directors.*

BOSTON, May 8, 1863.

LETTER FROM W. H. STEVENS.

—

APRIL 10, 1863.

L. W. CLARKE, ESQ.:

DEAR SIR,—Your esteemed favor of the 1st instant came duly to hand. My reply I herewith transmit for your consideration. The Star mine is situated two-and-a-half miles southeast of Copper Harbor, upon Sections Nine, Ten, and Fifteen, Township Fifty-eight North, of Range Twenty-eight West; containing six hundred acres; covering the principal metaliferous zones of Keweenaw Point.

The most important zone, and that which should command your earliest attention, is situated immediately south and underlying the green stone. Within a space of three thousand feet there are seven important cupriferous amygdaloidal beds, two of which are of much importance for the copper they contain, and all of them more or less so for their influence upon the transverse veins.

The first and most important work to be commenced, is the adit gallery, which should be driven as fast as possible until it reaches the green stone, a distance of from four hundred to five hundred feet, through productive ground. A shaft should also be sunk at the brink of the green stone to the adit gallery; depth of shaft, one hundred and eighty to two hundred feet.

A lift of pumps should be put into No. 1 shaft, say five-inch working barrel, six-inch pipe two hundred feet in length, which will dry the third level below the adit.

One of Bird's portable engines, of the same size you have now at the mine, would be sufficient to drive the pump, while the one on hand will be ample for hoisting for one or two years. At the same time you should improve the water power of the Montreal river for stamping purposes, and so erect the stamps that the adit tram road may be extended direct to the stamps.

I have estimated the power of the Montreal river on your estate, as follows: equal to 4 feet in width, 2 feet in depth; velocity, 150 feet per minute; gives 75,000 pounds; 30 feet fall gives 2,250,000 pounds; at 33,000 pounds, 1 foot per minute, gives 51 horse power, which will be equal to stamping 100 tons of your stamp mineral per day.

The principal vein is large, well defined, a true copper-bearing lode, varying from two to six feet in width, composed of calc, spar, quartz, chlorite, laumonite, with copper disseminated throughout. I should judge it to contain from two hundred and fifty to five hundred pounds of copper per cubic fathom.

I fully believe the more northern portion of the mine (as yet unopened,) will be more productive than where it is now opened; and from recent developments at the Central mine, and the promising appearance of your mine at the surface, we have good reason to expect more productive ground south. The promising appearance and productiveness of the amygdaloidal lodes at the surface, adds very much to the value of the mine, from the fact that they can be drained and wrought through the openings of the mine; and the openings, tramways, etc., can be so constructed that the mineral can be transported direct from the mine to the stamps without change, and with little handling.

Duly considering the proximity of a good harbor, good roads already constructed, an excellent growth of timber suitable for mining purposes and fuel, an excellent water power, and the advantages of openings and drainage, the improvements upon the premises, and above all, the rich mineral prospects, I cheerfully recommend it as a valuable mining estate, and will say that I know of no other mine in the district that can be opened so quickly, and with so small amount of capital, as the "Star."

Respectfully yours,

W. H. STEVENS.

LETTER FROM A. C. DAVIS.

Office of the Amygdaloid Mining Co.
of Lake Superior,
June 11, 1863.

L. W. Clarke, Esq.:

Dear Sir,—I visited the Star Mining Company's property last week, and from all I could see of the property, was favorably impressed with its value.

I could not get into the mine to examine the vein, but judging from the amount of work done, as shown by the mine plan and the amount of vein rock on the surface, I am led to believe the vein to be a large one, and well worthy of farther trial, as I understand there has been no stoping done in the backs, and that the openings are in the rocks, which are not the most productive points of the mineral range. I think, as your levels approach nearer the green stone, you will find the vein to carry heavier copper than you have yet seen.

The vein stone that I examined, I was informed, had been culled over for the barrel work, and had yielded a fair per centage of it. I think the product of this lode resembles the Connecticut vein matter more than any I have seen before. The copper is fine, but is still granular, and can be saved easily with careful washing, and will yield, I think, much better than its appearance indicates. The rock will stamp easy, and the greater portion of the vein being calcareous, is favorable for its separation from the copper.

I examined one of the belts of amygdaloid that has a small opening on it, and it is as well charged with heavy copper as any belt I have seen with the like openings. I think I have never examined a tract of land that has greater natural advantages for mining.

I deem the water power quite sufficient to drive forty-eight heads of stamps, and furnish abundance of waste for washing. Its value to the mine can hardly be estimated. The mine being only three miles from a good harbor, with a good wagon road already completed, is certainly much in its favor; and I think the tract well worthy of a good trial by actual mining. I understand there are other veins on the property, of much promise, as shown by surface indications.

Respectfully yours,

A. C. DAVIS.

REPORT OF D. S. CHILDS.

STAR MINE, JAN. 15, 1864.

H. W. NELSON, ESQ.,

Treasurer Star Copper Company, Boston, Mass.:

DEAR SIR,—In the last of May, 1863, I reached the Star mine, to re-open it, and commence mining again.

The work to be done, was to clear the mine of water; erect a stamp mill; build a dam across the Montreal river; build a canal, seven hundred feet from the dam to stamps; build a tram road from the mine to the stamps, etc.

We found the houses built during the old working of the mine, in a dilapidated condition, and it was thought well to repair them, as best we could, for present use. Yet those old buildings seemed to check the progress of our work in every branch; men would not stay with us, because they could have better accommodations at other places.

The Secretary, having learned this fact, sent me, late in the fall, an order to erect more comfortable dwellings; upon which order have been built the following: One dwelling for agency, 34 by 38 feet on the ground, 34 by 26 feet, 1½ stories high; one office, 1½ stories high;

two boarding houses, 30 by 32 feet on the ground, 20 by 30 feet, 1½ stories high,—these houses, built of hewn logs, make very comfortable dwellings; one frame house, lathed and plastered, 34 by 20 feet, with kitchen 10 by 14, 1½ stories high; one house for mining supplies and captain's office, 12 by 16 feet, 1½ stories high, a frame building; company's barn, also frame, 20 by 26 feet, 2 stories high; one log barn, for the contractor who teams our supplies, lumber, iron, castings, etc.

The opening of the mine was commenced under adverse circumstances. A small engine which had been formerly used here, we found in a very bad condition. Many of its fixtures having been taken away, we were obliged to supply others from Detroit. It was over one month before the repairs were completed so as to commence work.

The mine was full of water to the adit level; and having no pump, we drew off the water in barrels, which was a slow process. Below high water mark we found a large portion of the ground had caved. Large slides had taken place in the lode, choking every level in the mine, from shaft to shaft, excepting the adit level. The timber in the shafts, sumps, stulls, etc., being rotten and worthless, the whole mine had to be re-timbered, and new ladder-roads put in. We built a new shaft house at No. 1 shaft, as the old one was too low. The re-timbering of the shafts was a slow and dangerous work.

Twenty-five fathoms of ground have been stoped near and adjacent to No. 1 shaft. The adit level has been driven one hundred and forty-seven feet and seven inches north.

To the amount of these add that of stoping and drifting, from Mr. Hall's expense account for December.

The shaft on the belt of amygdaloid has been sunk thirty-four feet and nine inches. It is now idle, not for want of sufficient copper, although it was not very rich excepting at the surface; but when winter came the miners would not stay, because they could find work in warmer and deeper mines.

A shaft on the Clarke vein has been sunk seventeen feet six inches, and stopped for the same cause. The stamp work from these stopes has not been rich, and there was no barrel work. The adit level was opened in one hundred and twenty feet north of No. 2 shaft, when we re-commenced work. There was no sign of copper in it when we started; it has been driven as above stated, and now we have a vein twenty-two inches thick, fairly charged with copper. We have now a contract to drive it two hundred feet, which will take till the middle of April or first of May. This will bring us near the green stone, when, if the vein increases in richness, as in the last hundred feet, we shall have a very fair vein.

The dam is completed, with a waste wier in the

bottom on the bed of the stream, capable of discharging a large volume of water. A waste way is cut in the solid rock, on the south bank of the stream, to carry off the surplus above the level of water we want in the canal. We have a gate at the dam, to shut the water out of the canal at any time. The canal is completed to with one hundred and twenty-five feet of the stamps.

The stamps and wash house having been put up and enclosed, we are at present arranging for machinery inside. The stamp building is 34 by 39 feet; the wash house is 40 by 60 feet, all of the best frame work. The tram road is nearly completed, there being work to do only on the intermediate chutes. This road is nearly all on trestles, and has taken a great deal of timber. It stands clear of the ground, so that the snow, which retards the work on so many of the tram roads in this district, can be easily cleared away.

The teaming has mostly been done by contract, and at a much lower rate than we could have done it on Company account. During the summer we have had two yoke of oxen drawing timber for the dam, stamps, tram road, etc. Now that most of this work is done, hay being very scarce and of high price, I have disposed of one yoke.

As regards the future prosperity of the mine, I would say that I do not know of better mining property in this neighborhood. We have land enough for mining purposes, and timber sufficient for many years.

We have confined our mining, this year, to the old mine, which was abandoned long ago, because it would not pay. But I begin to feel confident that if we get under the green stone, we shall find more copper, even in the Star vein.

Since I came here, we have uncovered two well defined veins ; one lying seven hundred feet to the west of the Star vein ; and one six hundred feet to the east. The one west, for designation, I have called the Big Break vein. This we first opened one hundred feet from the green stone. At this point it is about five feet wide, and well charged with shot copper. We opened it again one thousand feet south, where there is not as much copper as nearer the green stone ; showing clearly that the nearer we approach this belt of rock, the richer we find the veins. It shows clearly that it is a regular vein. An easy method of proving the richness of this vein, at a depth of eighty feet, is to clear out an old shaft, eighty feet to the west, sunk in a small vein which produced quite an amount of copper,—so much that miners worked it willingly on tribute. A cross cut may be made from this shaft to the vein. A gallery might then be driven on it.

I believe, with Samuel W. Hill, W. H. Stevens, Dr. Clarke, and other mining men, that we should then have a richer vein than has ever been seen on this property. The expense of reaching the lode, on this plan, would be about twelve hundred dollars, at a depth of eighty feet from the surface. To sink a shaft this depth would cost seventeen hundred dollars. In case this vein is found good, a level could then be driven to the surface, on the course of the vein. Judging from its course, the shaft is sunk upon a small branch of the main, or Big Break, vein. Its being so rich is good evidence that the main lode to which it belongs, when reached at a depth where the rocks are settled, will be valuable.

A lode opened to the east of the Star vein is considered to be the Clarke vein, opened by a French company directly north of us. The small vein, on which the amygdaloid shaft was started, was supposed to be this vein.

This Clarke vein was uncovered in September last. It was really promising; and after opening it in five different places, a contract was given to sink a shaft upon it. Of this vein we have five hundred feet in length, south of the green stone, and as we go under the green stone, on the inclination of strata, the length will increase, There is sufficient length to open a good mine on this vein.

I am strongly impressed with the opinion that at the north of the green stone, on Section 10, we can open some veins which will prove of immense value. It is thought that good copper-bearing rocks on the Point, lie north of the green stone, rich veins having been opened on Section 7, by Mr. Stevens, the past season, which are considered as promising as anything in this district.

Also the Clarke mine, joining our lands on the west, has a very fine vein, and when worked again will probably prove profitable; and there is no reason why this property, having the same belts of rock, should not prove as good. The Star Company own a mile east and west. We know of three veins in the green stone strata, well defined and literally charged with copper. The rocks on the northern slope are near the surface, and hence are easily explored. As soon as snow is gone we shall open the veins which traverse the property, and if the explorations answer our expectations, before June, 1864, I shall have the pleasure of announcing to the Company a discovery of something worthy of note. I have no hesitation in saying that another summer will develope this property to a very desirable extent.

Very respectfully,

Your most obedient servant,

D. S. CHILDS,

Agent Star Copper Co.

www.ingramcontent.com/pod-product-compliance
Lightning Source LLC
LaVergne TN
LVHW011146110826
845150LV00008B/2548

* 9 7 8 1 4 1 8 1 9 2 2 6 6 *